Tracking

The Basics

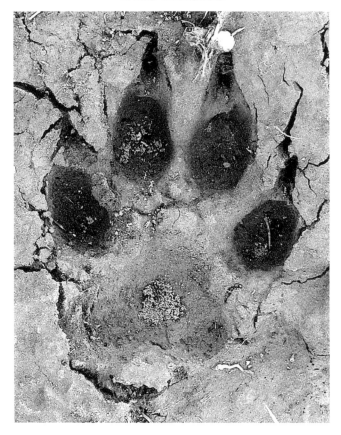

James C. Halfpenny, Ph.D.

Tracy D. Furman, B.A., M.A.T.

Published 2010 Copyright by James C. Halfpenny

All Rights Reserved. No Part of this book may be reproduced in any form or by any electronic or mechanical means, including information storage and retrieval systems, without written permission from James Halfpenny.

First Edition 2010   10 9 8 7 6 5 4 3 2 1 printing
Printed by Createspace.com   ISBN 1456306383   EAN 9781456306380

Published in the United States by
A Naturalist's World
www.tracknature.com --- (406) 848-9458
PO BOX 989, Gardiner, MT 59030

WOLF versus DOG CHECKLIST

- ✔ Track: size and shape
- ✔ Track: toe size and shape
- ✔ Track: splaying
- ✔ Track: claws show
- ✔ Gait: type
- ✔ Gait: stride
- ✔ Gait: straddle
- ✔ Trail: scent posts
- ✔ Trail: den site
- ✔ Scat: size and quantity
- ✔ Scat: content

- ✔ Season
- ✔ Behavior: wandering
- ✔ Behavior: lack of direct register
- ✔ Behavior: galloping and bounding
- ✔ Behavior: sloppy kills
- ✔ Behavior: feeding
- ✔ Behavior: returning to kill
- ✔ Behavior: caching food
- ✔ Human: proximity to towns
- ✔ Human: tracks
- ✔ Human: ski or dog sled trails

Acknowledgements

We wish to acknowledge Sarah Boles, Koani, Kerry Murphy, John Olson, Mike Phillips, Elizabeth Rogers, Doug Smith, Dan Stahler, David Tiller, Jon Trapp, Pam Troxell, Pat Tucker, Bruce Waide, Adrian Wydeven, Defenders of Wildlife, Yellowstone Association Institute, Yellowstone Wolf Project and all the instructors, students, participants and volunteers through the years for their help. Our apologies to those, who due to limits of space and frailty of mind, we have not been able to list.

Ordering Footprint Casts, Books, and Classes

Permanent plastic and cold-cast bronze footprint casts of wolf and other mammal tracks may be ordered through A Naturalist's World at
www.tracknature.com
on the internet. A list of our field classes and other publications will also be found there.

TABLE OF CONTENTS

Tracks: **Pages 4-13**

Gaits: **Pages 14-23**

Scent Marking: **Pages 24-25**

Scat: **Pages 26-27**

Carcass Analysis: **Pages 28-29**

Comparisons: **Pages 30-37**

Comparison *Description*

INTRODUCTION

The purpose of *Tracking Wolves: The Basics* is to aid in the identification of signs left by wolves. Correct identification is critical to proving presence or absence and to studying wolf behavior and ecology. Knowledge of footprints, trails, and other signs left by wolves prevents confusion with similar species including domestic dogs, coyotes, bobcats, cougars, and even bears. While variations in sign exist, specifics and generalities about signs provide evidence needed to interpret the stories left by wolves in their trails.

This book is divided into two parts: the description of signs left by wolves and comparisons of wolf signs with those of similar species. Key characteristics provide visual confirmation of species. Measurements and life-sized drawings provide scale when looking at footprints. Detailed charts list comparative differences allowing differentiation of similar species. Color indexed tabs facilitate quick access to information.

Biology: An understanding of wolf biology and the general yearly life cycle of wolves enhances interpretation of trails. While among individual, pack, season, and geographic variation occur, it is possible to define generalized behavior and a Wolf Year. The female reproductive cycle includes two phases: heat, in simply terms preparation for estrus (approximately December 1 through birth) and estrus, in simple terms preparation for shedding eggs (late January through early February). During heat, the alpha or top female is both dominant over pack members and possessive of her territory. She may mark by making raised leg urinations (RLU) and double urinating where her alpha male does. Males exhibit interest in her, sniffing her genital region. Heat is ten days long at the end of which eggs are shed. Increased receptivity leads to mating near the end of estrus.

On average five-six pups are born about 60 plus or minus three days later. Pups emerge from the den in about 21 days.

During the socialization development phase, pups learn behavioral patterns for life including establishment of emotional bonds for pack continuity, dominance relationships, and feeding and predation behaviors. During the juvenile development phase, pups associate killing with eating, initiate killing behavior, participate in hunts, and grow to adulthood.

After approximately two months, the pups may be moved to a rendezvous site where older wolves take turns watching the pups while others hunt. By September the pups are traveling with adults and in late fall they join hunts. During their second year, as yearlings, they become successful hunters.

It is important to remember that even the biggest wolf is small once in its life before it grows up. In general, pups are born weighing less than a pound, weigh 13 lbs in a month, about 20 lbs when moved to the rendezvous site, and 40 lbs when they start traveling with the pack, and 60 to

> HINT: measurements in this book are based mostly on Rocky Mountain wolves and will be smaller for Mexican wolves and those from the Great Lakes region but larger for wolves from Alaska and Canada

100 lbs when one year old. Adult Rocky Mountain females weigh about 90 lbs and adult males weigh 110 lbs.

Signs: Wolf signs covered here include footprints, trails, scat (feces), scent posts made with urine, and kill sites. Footprints are the most diagnostic sign used to confirm the presence of wolves. Track sizes and shapes vary between hind and front footprints and among genders and ages. Footprints can also look different depending on the surface where found; soft surfaces may look larger and melted snow may increase apparent track size. Trails vary dramatically depending on gait (walk, trot, gallop), speed, and behavior. Scat, also varies with food content (fiber and moisture) and the size of the animal. Wolf behavior creates tell-tale clues at territorial scent mark sites and predation kill sites.

Tracking Wolves: The Basics is your guide to the key characteristics for interpreting gray wolf (*Canis lupus*) sign. The reference material contained describes how to correctly measure footprints and trails and read the stories left by signs.

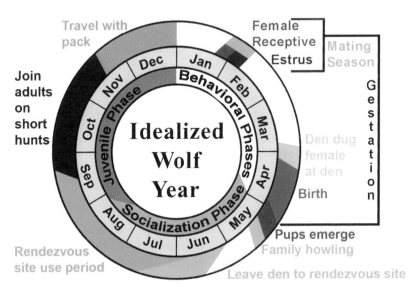

The idealized Wolf Year shows behaviors and their variability. Slanting lines roughly indicate when wolves start a behavior and when all have finished it (outside of line). Timing varies across the geographic range of wolves.

WOLF FOOTPRINTS

HINT: large domestic dogs may have tracks as large as wolves. See the comparisons section later in book.

PRIMARY CLUES
* footprint longer than wide
* four toes present
* claws usually show
* single front lobe on interdigital pad

SECONDARY CLUES
* front lobe is shallow
* three posterior lobes on interdigital pad
* toes symmetrical about center line

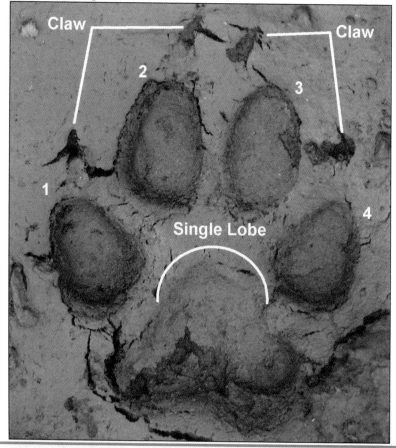

INFO: the interdigital pad is the large pad on the back of the footprint that extends forward between the toe pads.

FRONT AND HIND FOOTPRINTS

HINT: don't mistake front and hind footprints for two different animals as these footprints may vary dramatically.

PRIMARY CLUES
* front size larger than hind
* front interdigital pad larger
* front interdigital pad occupies proportionally more of footprint

SECONDARY CLUES
* front toes splay more
* front gap smaller
* front interdigital pad wings larger

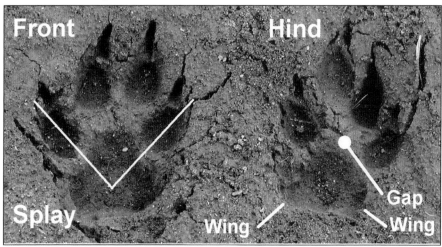

INFO: the interdigital pad is composed of three sub-pads, a medial and two laterals called the wings. The gap is the distance between the leading edge of the interdigital pad and the back of the closest center toe.

MEASURING FOOTPRINTS

HINT: proper measurements mean correct identification.

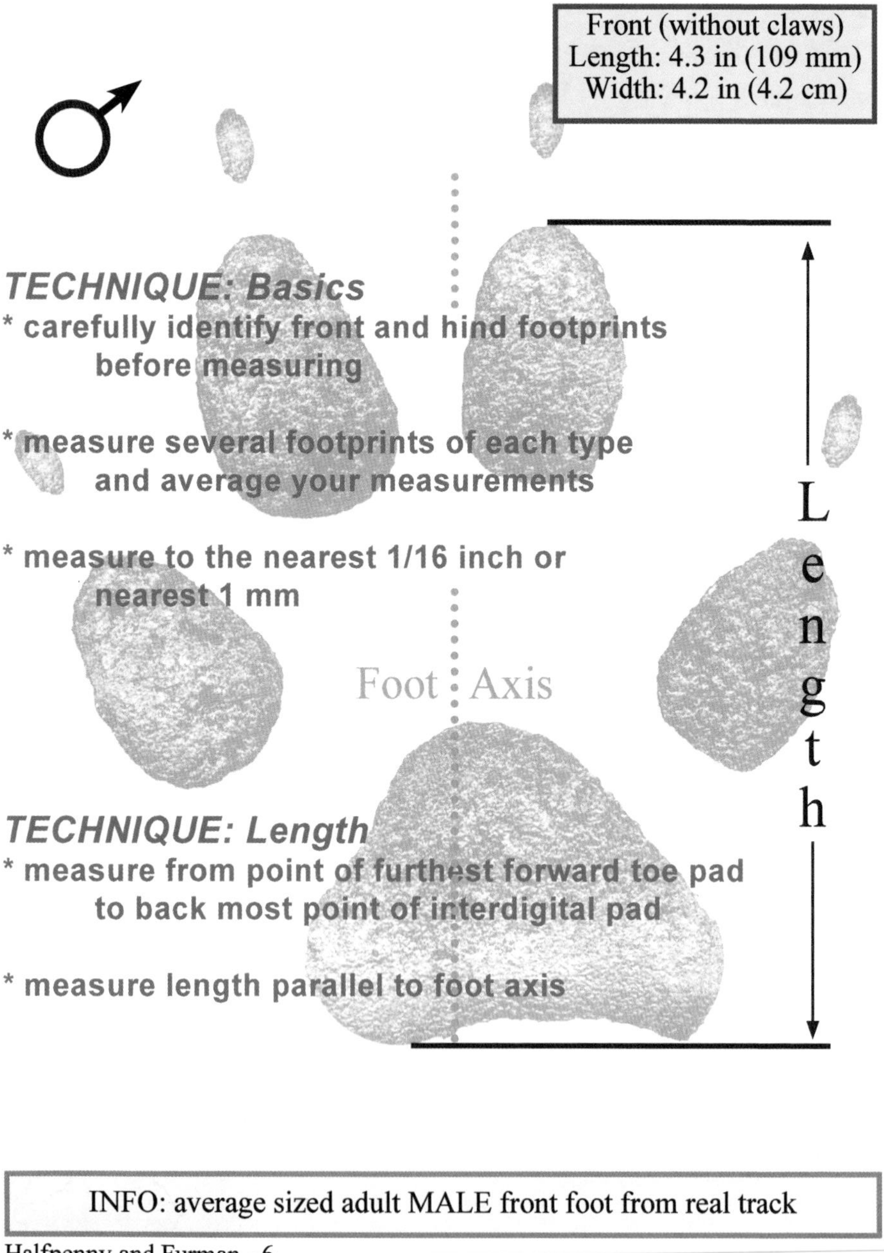

Front (without claws)
Length: 4.3 in (109 mm)
Width: 4.2 in (4.2 cm)

♂

TECHNIQUE: Basics
* carefully identify front and hind footprints before measuring

* measure several footprints of each type and average your measurements

* measure to the nearest 1/16 inch or nearest 1 mm

Foot Axis

Length

TECHNIQUE: Length
* measure from point of furthest forward toe pad to back most point of interdigital pad

* measure length parallel to foot axis

INFO: average sized adult MALE front foot from real track

Halfpenny and Furman - 6

MEASURING FOOTPRINTS

HINT: take good measurements or your numbers mean nothing.

Hind (without claws)
Length: 4.3 in (109 mm)
Width: 3.5 in (88 (mm)

TECHNIQUE: Width
* measure width perpendicular to foot axis

* measure from the furthest right point of an outside toe pad to the furthest left point of the other outside toe pad.

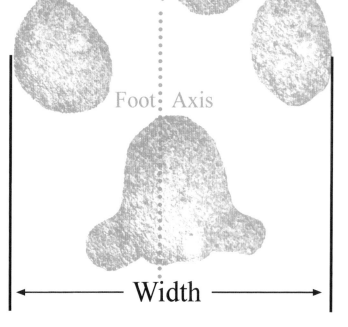

INFO: average sized adult MALE hind foot from real track

Tracking Wolves - 7

MEASURING MINIMUM OUTLINE

HINT: when a footprint sinks into a soft surface it appears bigger

Front (without claws)
Length: 4.1 in (104 mm)
Width: 3.6 in (92 mm)

TECHNIQUE

* a wet foot leaves a true footprint on a hard surface, such as concrete

* when it sinks into the surface, the footprint appears larger to the eye and is exaggerated (see diagram below)

* The exaggerated outline is known as the variable outline because the size varies with depth

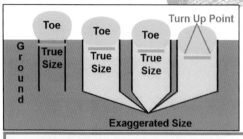

Cross-section (left) of ground showing a toe sinking in the surface to different depths. The minimum outline is the true size of the footprint.

INFO: average sized adult FEMALE front foot from real track

MEASURING MINIMUM OUTLINE

HINT: measure the minimum outline, the true footprint size

Hind (without claws)
Length: 3.9 in (99 mm)
Width: 3.4 in (87 mm)

TECHNIQUE

* look for where the edge of each pad turns up from a hard surface

* mark furthest points top, bottom, right, and left with a point made by a sharp object

* measure the true size using minimum outline points

INFO: average sized adult FEMALE hind foot from real track

Tracking Wolves - 9

MEASURING

HINT: note difference between true size and variable outline

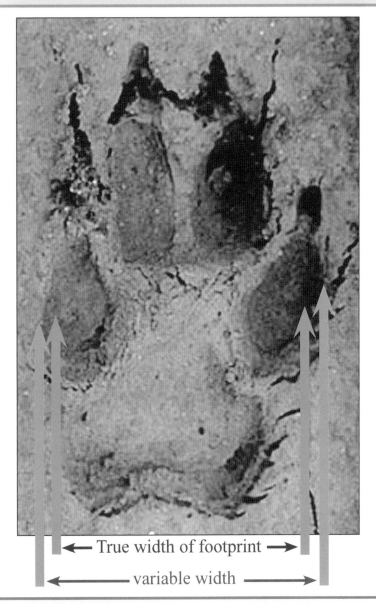

← True width of footprint →
← variable width →

INFO: minimum outline is independent of surface; same in all surfaces

FOOTPRINT SIZES

HINT: do not be fooled by footprint size variations

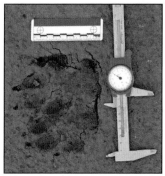

Considerable difference exists in the size of wolf prints because of age, gender, health, and regional variation. Minimum outline measurements reduce among observer variability and errors. Graphs below provide a visual look at amount of variation.

The top graph shows the minimum outline measurements of the entire footprint and bottom graph shows measurements of the heel (interdigital) pad. Brown diagonal line shows where length equals width. A dot above the line appears wide and below the line appears long. The greater the distance from line, the greater the impression of shape. Prints generally appear long while heel pads appear wide. Sample sizes (n) are shown.

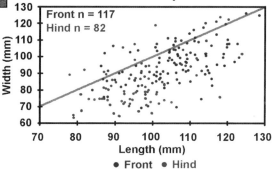

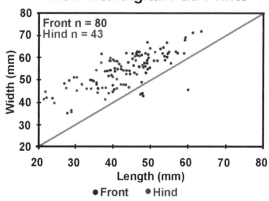

INFO: every wolf had a small footprint once in its life, some always do

AGE AND GENDER

HINT: cutoff lines, based on northwest wolves, are only suggestions

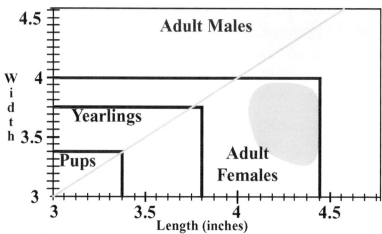

TECHNIQUE: Cutoff values
* compare length and width measurements of front feet to above graph
* adult male tracks generally measure in blue area
* adult female tracks generally measure in pink area
* yearling and pup tracks fall any where between cutoff lines
    (adults are considered to be older than 2 years)

CUTOFF VALUES
* Adult Males: larger than 4 $^{7}/_{16}$ x 4 in (112 x 100 mm)
* Adult Females: larger than 3 $^{13}/_{16}$ x 3 $^{3}/_{4}$ in (97 x 95 mm)
* Yearlings larger than 3 $^{3}/_{8}$ x 3 $^{3}/_{8}$ in (85 x 85 mm)
* Pups less than 3 $^{3}/_{8}$ x 3 $^{3}/_{8}$ in (85 x 85 mm)

TECHNIQUE: Length / width ratio
* male, especially adult, footprints often appear wide
* female footprints often appear long
* if the length divided by width is greater than 1.0
    suspect male
* when less than 1.0, the closer the ratio is to 1.0 the higher the
    probability that the animal is a male

INFO: cutoff diagrams are based only on front footprints

AGE AND WEIGHT

HINT: careful measurements of footprints reveal wolf age and weight

All mammals, including wolves, do grow rapidly from birth to adulthood. During the growth period it is possible to estimate age. To estimate the age, measure the front length or width in mm. Draw or visualize a line from the bottom axis to the curve. Then draw or visualize a line left to the vertical axis and an estimate of age in days.

When an animal becomes an adult, bone growth ceases and age estimates become unreliable. More calibration is needed for these growth curves.

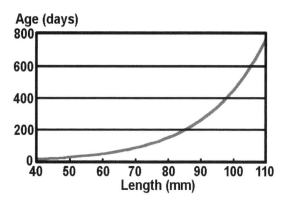

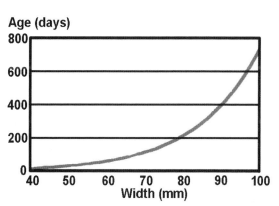

During the growth phase, reasonable estimates of wolf weight can also be made from front footprint measurements with the following equations:
 weight (lbs) = 4.799(heel pad width, mm) -109.257   [$r^2 = 0.80$]
 weight (lbs) = 1.875(print length, mm) -85.773   [$r^2 = 0.69$]
 weight (lbs) = 1.1516*e^(0.0464*(print length, mm))   [$r^2 = 0.72$]
These equations are based on a female wolf and may underestimate male weights. Footprints lengths longer than 100 mm and heel pads widths wider than 52 mm yield poor estimates of weight.

INFO: once bone growth stops, foot size only increases by adding fat

GAITS

> HINT: trails often reveal identification and behavior of the maker

Wolves travel great distances hunting and marking territories. They use a variety of gaits during their movements. These gaits can reveal information about who left a trail, how fast it was moving, and what it was doing.

As a generality, gaits progress from walks to trots to lopes to gallops. The progression provides more speed but costs increasing amounts of energy.

To understand gait patterns, key measurements are stride and group. The stride provides information about animal size and speed. When walking, stride is approximately equals the hip to shoulder distance of the body. A walking gait indicates size of an animal. In other gaits, the ratio of gait stride to walking stride defines relative speed.

A group (2 fronts, 2 hinds) defines and names the gait. After a group pattern, the animal may use any gait to start the next stride.

> INFO: gaits,sequences of body movements, leave identifiable patterns known as walks, trots, lopes, gallops, and bounds

GAIT MEASUREMENT

> HINT: length of walking stride equals mechanical size of wolf

Hip-to-shoulder length

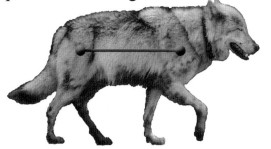

The walking stride approximates hip-to-shoulder length of a wolf. A 34-in (86-mm) stride indicates a wolf of about 34-in long. Do not confuse a walking with a trotting stride which will be more than 60 in (86 mm) for an adult.

Stride is measured from a point where a foot touches the ground to next spot where the same point on the same foot touches the ground again. The point may be any point that is visible in two successive footprints and may be either on a front or hind footprint. Stride is not measured across a change in gait patterns.

Amble

Walking Gait: smaller hind on larger (pink) front footprint

> INFO: stride and group measurements are parallel to the line of travel

Tracking Wolves - 15

WALKING PATTERN

HINT: determine the overall trail pattern and line of travel

Walking pattern: left / right footprints that are evenly spaced.

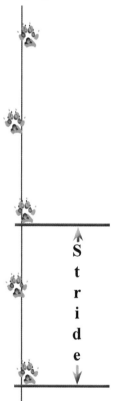

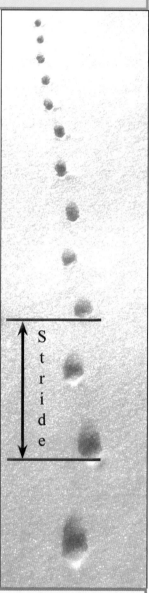

Walking stride of adult varies from 30 to 38 in (75 to 100 cm).

INFO: straddle, the width of the trail, of a wolf is narrow and insides of footprints may overlap. The trail of a deer or elk is wide; no overlap.

TROTTING PATTERN

> HINT: pattern right / left with even spacing, the gait is a walk or trot

Trotting pattern: left / right footprints that are evenly spaced.

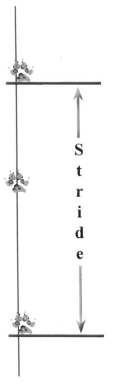

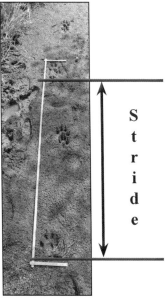

When trotting at a regular speed, hind foot registers directly on top of front footprint. At faster speeds, hind foot registers forward of the front footprint.

Trotting stride of adult varies from 60 to 76 in (150 to 200 cm) or longer.

Each segment of the ruler is 9.25 in (23.5 cm) long.

> INFO: stride is at least twice as long as hip to shoulder length

LOPE PATTERN

> HINT: groups of four footprints define lopes, gallops, bounds

Lope pattern: group of four prints separated from other groups.

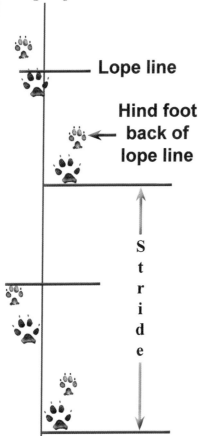

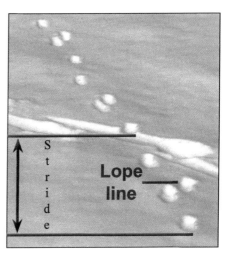

A lope is slow enough that a hind foot touches the lope line or even further back.

Lope strides vary from 4 to 8 ft (120 to 240 cm) and average 5.5 ft (168 cm).

> INFO: a lope is a gallop, simply a slow gallop

GALLOP PATTERN

HINT: identify front and hind prints to show this is a gallop not a lope

Gallop pattern: full speed gallop has hind feet forward of lope line.

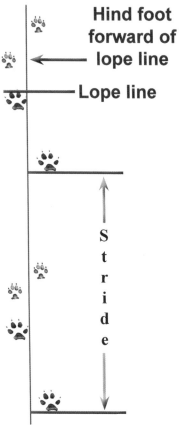

Hind foot forward of lope line

Lope line

Stride

Stride

Gallop stride may be 25 ft (760 cm) or more.

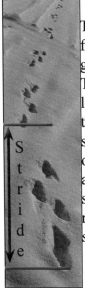

The groups of four footprints indicate gallops or bounds. The faster the gallop, the narrower the straddle (outside right of trail to outside left of trail) and at a very fast speed, footprints may register in a nearly straight line.

At a fast gallop telling right from left footprints may be difficult. Gallops may exhibit different patterns (see the next page). The top pattern is a transverse gallop and the bottom is a rotatory gallop.

Stride

Stride

INFO: a gallop, relatively speaking, may be more than 8 times faster than a walk (gallop stride of 25 (760) /walking stride of 3 ft (91 cm)

GALLOP PATTERNS

HINT: four gallop patterns exist, depending on front and hind foot leads

"Z-" and reverse "Z-patterns" are called transverse gallops.

"C-" and reverse "C-patterns" are called rotatory gallops.

INFO: different leads may lend themselves to turns or stability, but if the stride is the same, relative speed is the same

Halfpenny and Furman

BOUND PATTERN

> HINT: hind feet placed side-by-side define a bound

Bound pattern: hind feet are used simultaneously and placed side-by-side.

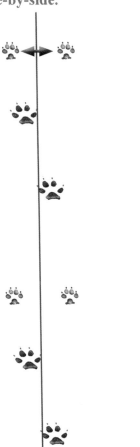

Bound strides may reach 25 ft (760 mm) but are generally shorter.

When bounding, feet are used simultaneously and placed side-by-side. Simultaneous thrust from the two powerful hind leg muscles provides great vertical height (jumping) or great horizontal distance. Using the power hind legs simultaneously also provides a solid base from which to bite a larger prey.

Bounds are called jumps when used for vertical height but bounds when used for horizontal distance. If a rabbit or rodent bounds, we call it a hop.

> INFO: whether vertical jumps, or horizontal bounds or hops, the pattern is the same with hind feet place side-by-side

SPECIAL PATTERNS

HINT: some gait patterns may reveal health, body position or behavior

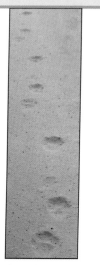

Injuries show in the trails of wolves. The animal on the left is using a transverse gallop with a badly injured left hind leg. It is not able to put its weight on the leg. Although injured, this wolf was still able to take down elk on its own.

The trail in the middle is from a wandering wolf. Wolves tend to wander more when young, during the breeding season, or simply when curious. Wandering is not a definitive clue of a domestic dog as feral dogs may travel as straight along their route as any wolf.

Canids, including wolves, often turn their bodies sideways (right) when traveling, called a "dog walk" or "dog trot." Positioning the head to the side allows the wolf to see over 180° including where it is going, the rest of the pack, and what is behind. The view is limited to one side with the gait pattern indicating where the wolf's attention was focused. Location of the larger front footprint shows where the wolf was looking. In the trail, the wolf was looking right. All gaits may be turned to the side and are called side gaits, for example, side walk, side trot, and side lope.

INFO: front prints are larger, more robust, with bigger pads than hind

READING GAIT PATTERNS

> HINT: great trackers understand the stories told by gait patterns

To understand how a wolf is moving read its gait pattern. Gaits reveal much about behavior and what is "on a wolf's mind." If a wolf changes gaits, good trackers look for the reason why. It may simply change gaits but often there is a specific detectable reason.

Walking: Wolves are active animals and seldom walk. A walking wolf is often showing caution such as when approaching a strange object. If the wolf is coming up-wind then it is nervous about what it is approaching.

Trotting: Wolves prefer two gaits: trotting and a C-shaped rotatory lope. When trotting a direct registry indicates a regular gait with no urgency. As the wolf moves faster the hind footprint moves forward of the front (similar to a walking amble). Greater distance of the hind foot forward of front indicates increasing levels of urgency or a young wolf trying to match the stride of an adult.

Loping: By virtue of the fact that a lope is a slow gallop, it shows that there is no urgency of actions. The C-shaped rotatory lope and the trot both are trade-offs. Using either of these gaits allows the wolf to cover more ground in search of prey, but does not cost the wolf too much energy. Going too fast would use more energy than provided by food procured. Loping wolves often slow to rest and a tracker may catch up.

Galloping: Gallops are gaits of urgency and speed. A galloping wolf may be closing the distance to a prey animal, fleeing an interaction with another animal including a human, chasing a rival during breeding, or chasing an intruding pack member from its territory. If you are following a wolf that has elected to gallop, odds are you will not catch up with it.

Bounding: Bounds provide force for close distance grabbing of prey, jumping vertically, or leaping across or above objects such as streams and fences. Bounds are used during the final phase of chasing prey. The side-by-side placement of feet provides a platform allowing a smaller wolf to make a solid grab at a larger prey; to pull down an elk by the throat.

Speed: Relative speed is indicated by stride. Divide the stride, in feet, by 3 (approximate walking stride of an adult wolf) to estimate relative speed. For example a 9-foot stride means the wolf is moving 3 times faster than a walk. Caution, do not measure stride between different gaits.

> INFO: 4 gaits (walk, trot, gallop, bound) reveal most wolf behavior

SCENT MARKING

HINT: urine is used to mark wolf territories for boundary defense

The elevated position of the urine on the snow on the conifer branch indicates a raised leg urination (RLU) by a dominant or alpha male. Urine icicles formed when urine dropped to the ground. Note territorial claw scratching.

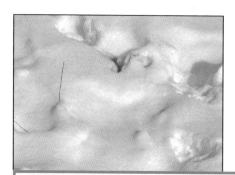

The position of the urine stain in the center of the trail indicates a subordinant wolf. The linear urine streaks suggests a male.

INFO: urine marks in the trail are made by subordinant wolves

SCENT MARKING

> HINT: blood in urine may be due to estrous or an internal injury

Two urine stains indicate two wolves or two passes by a single animal. The elevated position of the urine indicates raised leg urinations made by two dominant or alpha animals, called a double RLU.

The position of the urine stain to the side of the trail indicates a dominant wolf. The linear urine streaks suggests a male.

> INFO: urine marks to the side of the trail are territorial

SCAT IDENTIFICATION

> HINT: scat are used to mark territories and identify individuals

Points on the end of scat cords (top and right) suggest the makers were canids. Scat composed of animal protein is dark brown to black in color. Gray colored scat often contains bones and hair

This territorial scent post (left) has scat representing several visits at different times.

The older a carcass, the drier the meat and the more segmented the scat becomes. Dog food often contains grain filler and is very granular.

Wolves will eat grass perhaps to scrape parasites out of the intestinal tract

> INFO: scat are not as definitive for identification as are footprints

SCAT SIZES

> HINT: scat diameter provides a clue to species identification

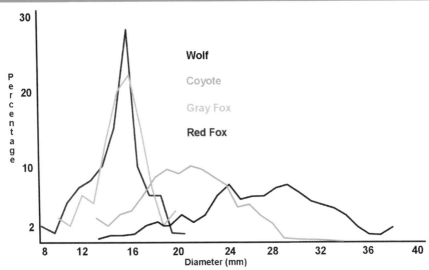

While larger canids have larger scat, there is considerable overlap in diameter (above). Total quantity of scat provides an additional clue with the quantity of wolf scat being large. When coyotes eat the same food, perhaps at the same carcass, their scat will appear just the same as wolf scat and size may help differentiate the two species (below). Even cats eating fresh meat will produce scat that looks like that of a canid.

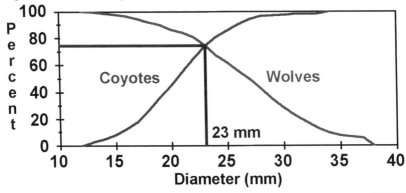

> INFO: when identifying scat, consider both size and quantity

CARCASS ANALYSIS

> HINT: on the predation trail find chase, kill, and feeding locations

With wolves, the chase portion of the trail is often long and may involve stalking. The kill site, often identifiable by disturbed ground, covered with intestinal and stomach materials, may be separated by a considerable distance from the feeding site.

Wolves mouths are small relative to large prey. They typically bite five areas: throat, front and back of front and hind legs. Kill bites are usually to the throat (note entry at throat bite on right) where the wolves hang on until prey dies. Wolf kills tend to be efficient and clean, lacking ripped skin and multiple bites. However, young wolves may not be efficient and sometimes do not finish the kill but leave livestock injured.

Entry to a carcass is often gained through the initial wound site and may be at the throat, or lower at the back portion of the abdomen. Carcasses usually are dragged, leaving the intestinal tract to trail behind, and exposing internal organs for feeding.

> INFO: kill and feeding methods are variable, not totally diagnostic

CARCASS ANALYSIS

> HINT: necropsy prey to determine if it was killed or scavenged

Wolves tend to crumple skin in an accordion fashion creating several large pieces. Edges of the skin are torn by multiple bites and thus frayed. Heart, liver, and lungs are eaten and large bones may be scattered. Small and medium bones may be chewed and frayed.

Center distance between upper canine teeth averages 1.5-2.25 in (32-38 mm) and lower averages 1.25-2.0 in (51-57 mm). Hole size and distance may be stretched in a long struggle. It is best to measure tooth distance under the skin during necropsy. Be careful when measuring multiple bites to match proper holes.

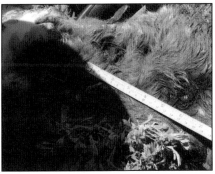

Additional clues to wolf predation include disemboweling during chase, bedding near the kill to digest, short, narrow trails to bed sites, occasional multiple kills, and returning to finalize a kill of previously wounded animal. Kills are marked with scat and urine. Wolves cache uneaten meat by burying, but also may scavenge carrion.

A necropsy (left) reveals hemorrhage indicative of live predation. Bone marrow reveals health of prey. Red jelly-like fatless marrow (left) shows poor health.

> INFO: best predation analysis requires immediate response

COYOTE SIGNS

HINT: coyote feeding sites are messy

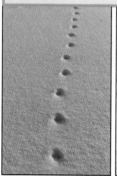

Coyotes mostly walk and trot with hind feet registering directly on top of front feet. Note the narrow straddle with insides of footprints overlapping (left). A side trot is shown on right; the coyote is looking right. Below on left is hind print and on right is front.

Coyotes do not move large prey from kill site. Instead they feed where the animal died; note stomach materials in background (below).

Coyotes seldom tear skin but pull mouthfuls of hair from the carcass to get at meat. Rib tips are chewed but no large canine teeth marks.

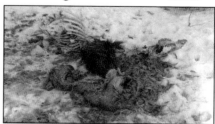

The diameter of 2/3 of coyote scats is less than 1 inch. Piles are small compared to wolves. Dry scat may segment and fall apart. Tapered tails are common. Coyotes are omnivores and scat may contain plants and insects.

Additional clues include bites to back of head, re-gripping throat, eating large prey at kill site, scattering bones, ragged skin and tendon edges, entering behind ribs to heart, and caching by burial.

INFO: coyotes prey mostly on small species and scavenge larger ones

BEAR SIGNS

HINT: bears leave big sign: tracks, scat, beds

Hind feet of bears have a full length heel differentiating hind from the shorter front feet. Right side of page shows an amble and grass and apple scat. Scat diameter is large (greater than 1.25 inches) and often with blunt ends. Total quantity of bear scat is large compared to canid or felid scat.

Carcass often opened vertically and heart, lungs, and liver eaten (left).

Large diameter canine tooth marks show at tips of chewed cow ribs (right) and skulls may be bitten through. The skin is often turned inside out and licked fairly clean.

Additional clues include short chase or ambush, kill with swat, crushing weight or bite neck/head, ground damage, broad trails from return visits, beds and scat nearby, moved / buried carcass, and protectively remaining near.

INFO: bears are mostly herbivores and diet may be 70% plants

BOBCAT SIGNS

HINT: bobcat and cougar sign are similar, differing mainly by size

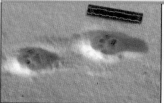  

The hind footprint of a cat (top left), is smaller but shape appears longer than wide compared to wider/rounder appearing front print. Cats usually walk with hind foot directly registering on front (top, middle) and a narrow straddle with inside of feet overlapping. Crossing openings, cats amble (right).

Habitat provides a good clue to cat trails. Cats tend to move stealthily using underbrush for visual cover. Cliffs provide shelter and additional cover. Cats will walk on downed trees, especially to avoid deep snow where their narrow feet sink in.

 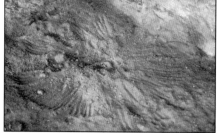

Cat scat is dry and often segmented. Ends are usually blunt.

Cats usually bury their scat by scraping loose soil over it.

INFO: dominant male cats often mark presence with unburied scat

COUGAR SIGNS

HINT: think like a domestic cat to understand bobcat and cougar

Adult cougar tracks are large and lack claw imprints. The front foot appears wide or roundish while the hind footprint (bottom print on right) is longer than wide. Crossing an opening, the cougar ambles with hind foot overstepping the front footprint.

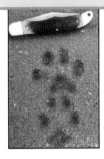

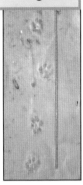

The method a cougar uses to kill prey depends on its size. A cougar bit through the top of the skull of a small coyote (left). Large prey are most often killed by biting the throat and holding on until the prey suffocates. Generally, cats do not have to re-grip a bite.

Cats have a rasp-like tongue surface which they use to scrape bone clean of all meat. Carcasses are buried with whatever materials are available. Cats revisit a carcass for up to a week, often moving it short distances each time.

Additional clues include short chases and ambushes, claw marks on carcass, move carcass with its intestines to seclusion for feeding, scraping skin clean, eating throat/shoulders first then viscera.

INFO: cats cheek mark trees, look for bark on ground, hair on tree

Tracking Wolves - 33

COYOTE DIFFERENTIATION

HINT: coyotes seldom move in packs; several members may spread out

Two species of canids pose difficulties when trying to separate their sign from those left by wolves: coyotes and domestic dogs. The next four pages explain how to differentiate coyote and dog sign from wolf sign.

Minimum outline measurements of adult coyote and wolf footprints effectively separate the two species. For front prints (right, large adult alpha male) a length of 3 and a width of 2.75 in (mm) is adequate. For hind prints a length of 3 (75 mm) and a width of 2.5 in (64 mm) is adequate. These measurements are more appropriate for western animals. For eastern animals use these measurements as a guide only as a detailed study is needed.

Several clues are helpful to separate young wolf sign from a large alpha male coyote. Expect wolf pups to remain at a rendezvous site until they are at least 120 to even 175 days old (adult signs will probably be visible). Pups quickly grow large, robust feet and by 60 days of age, wolf pup footprints are larger than coyote prints. A good set of clues is the width of toes: wolf inside toes are wider than .59 (15), outside toes wider than .47 (12), and heel pad wider than 1.5 (38) inches (mm) [please note the caveat about eastern populations above]. In additional, adult coyote toes tend to be narrow and pointed at the tips, whereas wolf pup toes are broader and more rounded as they have not grown into their adult shape yet.

INFO: pups tracks are nearly adult size by January

Halfpenny and Furman - 34

COYOTE DIFFERENTIATION

> HINT: coyotes spend a lot of time trotting, especially using a side trot

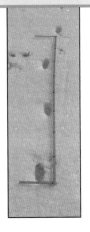 

The smaller coyote has a walking stride of about 17 in (43 mm) and a direct register trot of 34 in (86 mm) (left). Adult wolf strides for a walk start at about 30-36 in (75-90 cm) and reach 5 ft (150 cm) when trotting. The straddle (outside left of trail to outside right) of a coyote is narrow and the inside edges of the footprints often overlap (right).

Scat may be segmented, especially if dry, and may have a tapered tail both of which are also characteristics of wolf scat (right). The diameter of coyote scat which is mostly less than 0.91 in (23 mm) and the total quantity of coyote scat are both small compared to wolf scat.

Blood in coyote urine (above) is a sign of the reproductive season (estrus) but may also result from an injury. Blood is not a sign separating wolf and coyote urine.

Coyotes tend to scent mark (left) every $1/8$ - $1/4$ mi (200-400 m) while wolves tend to sent mark every $1/4$ to $3/8$ m (400-600 m).

> INFO: pure meat scat is black, fur and skin add brown, bones add white

DOG DIFFERENTIATION

> HINT: prints of dogs reflect their domesticated behavior and anatomy

In spite of what you may have heard or read, there are no clue or clues, either anatomical or behavior, that will separate the signs of all dogs from all wild canids. People have so altered the dog genome that any clue found in a wild canid is also present in some dog breed. An additional complication is that every wolf is small once in its life and some never grow large. The good tracker studies every clue and uses the totality of clues to make a final differentiation. Even though there are over 400 breeds of dogs, there are some generalizations that provide clues to separate dog from wolf tracks.

All four toes of dog prints often splay whereas with wolves if toes splay, it tends to be outer toes only. Claws tend not to register in footprints of large dogs.

Dog prints often tend to have toes that are disproportionately large compared to those of wolves, especially the center toes.

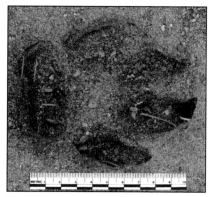

Large dog footprints tend to have a round shape whereas wolf tracks are more rectangular with length longer than width.

> INFO: dogs with medium-sized feet include Huskies, Malamutes, collies and German Shepherds

DOG DIFFERENTIATION

> HINT: feral or semi-feral dogs behave as wild, wolves do

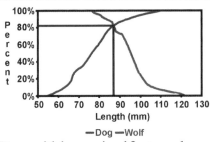
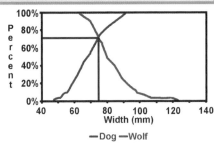

Dogs with large-sized feet may have footprints as large as the largest wolves. Nevertheless footprint size does provide a general guide for separation. When measuring a hind footprint a minimum outline cutoff of 3 7/16 (87) for length (top left) or 3.0 (74) inches (mm) for width (top right) separates wolf prints from dogs with the highest percentage of correct species, 82 and 72, respectively.

In addition to anatomical clues of size and shape, behavioral clues of dogs may include actions such as wandering trails and sloppy attacks on animals. Proximity to human activity is a key indicator. Use the totality of clues to separate wolf from domestic dog trails.

Widely splayed toes on the galloping trail on the left suggest a domestic dog. The trail on the right is a side trot. Splayed toes and human footprints suggest it was made by a domestic dog.

> See **Wolf versus Dog Checklist** in red, opposite the Table of Contents

> INFO: dogs with large-sized feet include Irish Wolfhounds, Akitas, Saint Bernards, Newfoundlands, Great Danes, and Mastiffs

Made in the USA
Charleston, SC
18 April 2011